MATH JOURNALING ADVENTURES

Logbook DELTA

LEARN THE SECRETS OF CREATIVE MATHEMATICS WITH DENISE GASKINS

This logbook belongs to:

_____

Year:

_____

Writing is how we think our way into a subject and make it our own.

Writing enables us to find out what we know, and what we don't know.

—William Zinsser, Writing to Learn

TABLETOP ACADEMY PRESS

Contents

Your Mathematical Adventure 4	Neutron Game 50
Join the Math Rebellion 5	Area Puzzle 52
Character Sheet 6	Julia Robinson Math Festival 54
David Harold Blackwell 7	Candy Puzzle 56
The Duel Puzzle 8	My Secret Rules 58
Warm Up Your Mental Muscles 10	Number Yoga 60
Jenna Laib's Counting Game 12	2-D Nim Game 62
Make a Pattern 14	A Generous Gift 64
Games 16	Fraction Pictures 66
Math Riddles 18	Collecting Data 68
Double Trouble 20	Connect the Dots 1 70
Perimeter Puzzle 22	Take-Back-Toe Game 72
Open Middle 24	Seeing Stars 75
Bus Puzzles 26	Fictional Math 76
The Substitution Game: Round 1 ... 28	How Crazy Can You Make It? ... 78
Round Rhombus 31	Math Eyes 80
Malcolm Swan's Partition Products ... 32	Pig Game 82
On This Day in Math 34	Stump an Adult 84
Pixel Graphics 36	Mini-Biography 86
Explain a Problem 38	Aliquot Game 88
Numberstorming 40	Flipping Pancakes 90
Hieroglyphic Numbers 42	Reinvent Your Homework 1 92
Half Plus Three 44	Sci-Fi Puzzle 1 94
Analogy 46	Can You Solve It? 96
Sorting Collections 48	Growth Mindset 98

Copyright © 2025 Denise Gaskins. All rights reserved.
Tabletop Academy Press, Boody, IL, USA, TabletopAcademyPress.com

About Math Websites: The Internet overflows with a wealth of math resources, but nothing is forever. If a website disappears, you can run a browser search for the author's name or article title. Or try entering the web address at the Internet Archive Wayback Machine (archive.org).

Contents

- Quarter the Square 100
- Which One Doesn't Belong? 102
- Create Your Own WODB 104
- What Would You Choose? 106
- Math Book Review 108
- Hexashapes 111
- Contig Game 112
- Explain a Puzzle 114
- Sci-Fi Puzzle 2 116
- Reiner Knizia's Banker 118
- Ratio Puzzles 120
- Connect the Dots 2 122
- Make a Million 124
- In the News 126
- Words Help Us Think 128
- Mountain Ranges 130
- Shannon Switching Game 132
- John Golden's Fraction Square 134
- Visual Patterns 136
- Create Your Own Pattern 138
- The Adventure of Learning 140
- Infinite Series 143
- Lines on a Grid 144
- Blockout Game 146
- Museum of Math 148
- Exponential Halves 150
- Make It Visual 152
- Hexangles 154
- US Presidents 156
- Math Art Challenge 158
- That's Mean 160
- Explain How 162
- Mental Math Workout 164
- Secret Number Codes 166
- Same but Different 168
- Cut and Paste 171
- Exponential Folds 172
- Math Poetry: Haiku or Senryū 174
- All Twos 176
- Math Pickle 178
- The Power of a Pattern 180
- Brain Dump 182
- Octagons 184
- Math Report 186
- Frayer Model 188
- Monthly Math 190
- Place Value Mastermind 192
- Math Riddles Redux 194
- Debate with Sofya Kovalevskaya 196
- Reinvent Your Homework 2 198
- Captain's Log 200
- Classify the Aliens 202
- Urbanization Game 204
- Strategic Thinking 206
- Don Steward's Rooftops 208
- The Substitution Game: Round 2 210
- Rear-View Mirror 212
- Special Thanks 214
- Discover the World of Math 215

Your Mathematical Adventure

Welcome to the team of math adventurers—humans throughout history who have joined together to explore the wonder of numbers, shapes, and patterns.

Most people don't realize that math is a social endeavor.

From the first artist scratching geometric designs in the sand to modern mathematicians expanding the frontiers of knowledge, people have always learned best when we learn together, sharing ideas and building on each other's thoughts.

How to Use This Book

This logbook offers more than 100 puzzles, games, investigations, and conundrums—prompts that lead you to discover new ways of looking at math or to rethink ideas you thought you'd mastered.

For some activities, you may not need the whole two-page spread. When you have extra space, you can:

♦ Try to find another, different way to answer the prompt. Sometimes a new approach leads to greater insight.

♦ Fill the page with a geometric doodle or a pattern of your own design.

♦ Write math rebel answers to your homework problems.

Other times, you may discover you need more space because your ideas flow beyond the confines of this book.

A Natural Cycle of Learning

There are no "right" answers. Instead, each prompt invites you to notice, wonder, and create your own math:

♦ *Notice* means to open your eyes and pay attention to details, examining all aspects of the situation, seeing beyond just the surface level. Write down everything you observe.

♦ *Wonder* means to respond to the things you notice, searching out relationships and connections to other concepts, diving deeper into the sea of ideas. Write down any questions you think of. Don't worry about answers, just brainstorm.

♦ *Create* means to process the things you notice and wonder, shaping them through your own perspective on the world, to make something new.

You might create an explanation, a story or poem, a drawing, a new question to investigate (mathematicians love finding new questions!), or whatever fits your own unique inspiration.

Like other math explorers through the ages, you may find this process easier and more enjoyable if you share it with friends. In real math, working together is never "cheating." It's the natural way for humans to learn and grow.

May you always enjoy the adventure!

Join the Math Rebellion

Math rebels write *any* true answer *except* what the textbook expects.

For example, if the textbook answer is 57, a rebel might write:

$$100 - 43$$

$$\text{Or } {}^{120}/_2 + (-3)$$

Or *"The total number of mushrooms in the basket, if three hobbits each picked nineteen 'shrooms (not counting the ones they ate)."*

Math rebels can make the answer as crazy as they like. Have fun!

Understand the Math

When you get a math worksheet or homework page, don't start working straight away. First, review the page to see if the problems look familiar. Do you know what the teacher or textbook wants you to do?

Math rebels always care about the truth. So first, learn what the problem means and how to figure it out. After you know how to solve the problem, then you can start working on your creative answer.

Choose Your Battleground

Fighting for intellectual freedom takes energy. Are you going to mess with just a few of the problems? Or will you turn the entire lesson into a protest statement?

Math rebels know the importance of justification. So be ready to defend whatever you write.

Another Way to Play

Instead of doing a long page of math homework problems on a single topic, follow the natural cycle of learning.

Work the first few problems, until you notice something that makes you wonder. Then spend the rest of your math time investigating that question.

You may not discover the answer to your question, but the adventure of exploring how math works will build a deeper understanding of your topic than a whole page of homework problems done by rote.

Live by the Two Rules

There are only two important rules in mathematics:

♦ You are allowed to write anything that makes sense.

♦ You are not allowed to write anything that doesn't make sense.

Anything else is just advice. Follow it if you wish, or blaze your own path.

Fight for Truth, Justice, and Creative Reasoning

Character Sheet

Fill this page with information and data about yourself. For example, you might include your height, age, the number of people or pets in your family, your birthday, favorite number or shape, history ("5 years ago I ...") or future ("In 3 years, I will ...").

CHALLENGE: Write the facts Math Rebel style, using an expression in place of each number. For example: "I have $8/4$ dogs."

David Harold Blackwell

(1919–2010)

David Blackwell studied probability, game theory, dynamic programming, and statistics.

He was elected president of the Institute of Mathematical Statistics and was the first Black American inducted into the National Academy of Sciences. Blackwell was also a prolific writer. He wrote one of the first textbooks on Bayesian statistics and more than 90 other books and papers.

He enjoyed thinking about statistical problems such as the Duel Puzzle (see page 8). In game theory, duels are a metaphor for any form of social or political competiton where only one party can win.

In 1967, Blackwell published one of his favorite proofs, linking game theory and topology: "It gave me real joy, connecting these two fields that had not been previously connected."

*David Blackwell,
Math Team Delta's Leader Emeritus*

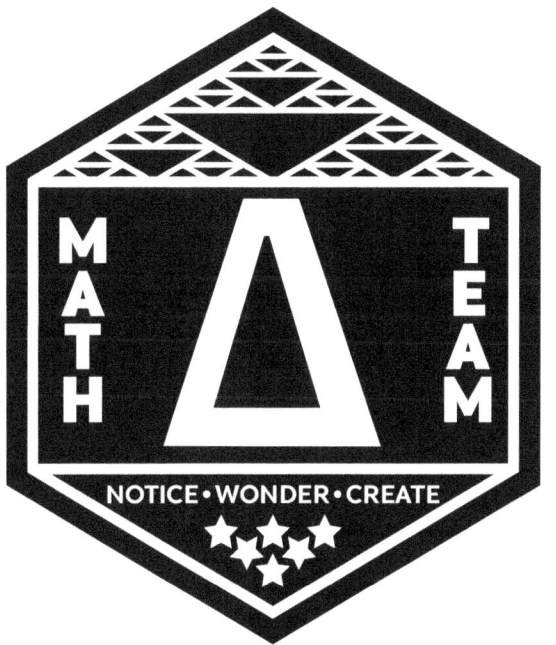

What Is "Delta"?

Mathematicians use many letters of the Greek alphabet to stand for different constants, variables, functions, etc.

The capital letter *delta* looks like a triangle. We use it to indicate the change in an amount.

For example, if you move from one point on a graph to another, "Δx" means "how much the x-coordinate changed."

The Duel Puzzle

Player1 and Player2 face each other. Both have a pistol, each with a single bullet—or paint ball, if you prefer. They alternate turns.

On each turn, they must either (1) take one step forward, or (2) fire a shot.

The closer the players get, the more likely their shot will hit their opponent. After they fire, they must continue walking forward, unless both players have already missed.

How might a player decide when is the best time to try a shot?

Warm Up Your Mental Muscles

Practice your Math Rebel skills. Pick any number, and see how many different ways you can write it. Start with a simple expression like addition or multiplication. Write an equal sign and another expression. Think about ways you can modify each expression to create a new one, and keep going until you fill the page or run out of ideas. What kind of fancy math will you create?

Jenna Laib's Counting Game

(two players or small group)

Choose a whole number, fraction, decimal, or negative number to skip-count by. Choose a starting place and a target number. For example, "Count by threes from 11 to 47."

Each player in turn chooses to write one, two, or three skips of the counting-by number. For example, the first player may write "11, 14." The second player might add "17," and then the next move could be "20, 23, 26." And so on.

The player who reaches or passes the target number wins.

Count by _____

Start at _____

Target = _____

EXTENSION: Mathematicians love to tweak games just to see what happens. How would you modify this game? Test out your new rules with a friend.

Make a Pattern

Create a pattern of numbers, shapes, or colors. Keep it going as long as you can. Will your pattern go all the way around the page? Or begin at the center and spiral outward? Or make any other design you like.

Need ideas? Do a search for "frieze patterns."

Games

Answer the prompt sentence. And then keep writing until you run out of room. Don't overthink it, just write. Keep your pencil moving. If you can't think of what to write, copy your previous sentence over and over until your mind comes up with something new to say.

What kind of games do you prefer, chance or strategy? Explain why.

Math Riddles

(any number of players)

Choose a secret number the other players will try to guess. Write a "What Number Am I?" riddle.

For example, "I am odd and prime. I'm a two-digit number less than 30. The sum of my digits is 4. What number am I?"

Give at least three clues for your mystery number. No other number should match all the clues.

EXTENSION: For more sample riddles, go to solveme.edc.org and click "Who Am I?"

Double Trouble

Write any number. Then write its double. Then double that new number. Keep on doubling. How high can you go before you run out of room?

Pick another number to double, and look for patterns. What do you notice? Does it make you wonder?

248

Perimeter Puzzle

A rectangle has a perimeter of _____ grid units. [Choose any number.] What might the area be? How many different rectangles can you find with that perimeter? What if the sides don't have to be whole unit lengths?

CHALLENGE: Perimeter values less than 4 units force the use of at least one fraction or decimal side length.

Open Middle

Go to openmiddle.com and click on your grade level, or just browse the topics. Find a puzzle you like. Copy it in your journal. Explain how you figured it out. Then make up an open middle puzzle of your own.

Bus Puzzles

A bus can hold _____ people. It starts out empty (except for the driver). At the first stop, _____ people get on. At the next stop...

Write a story for the bus.

What math questions might you ask about your story?

The Substitution Game: Round 1

(solitaire or small group)

Write a simple equation at the top of your paper. On your turn, write on a new line. Copy the equation from the line above, except replace any single number with an equivalent expression. For example:

$$2 + 5 = 7$$

$$2 + 8 - 3 = 7$$

$$2 + 8 - 3 = 14 \div 2$$

$$2 + 8 - 3 = (100 - 86) \div 2, \text{ etc.}$$

Before you leave, copy your final equation at the top of page 210.

Round Rhombus

Look at the math-art design. What can you say about the shapes or angles? Make a list of the things you notice. What do you wonder? Color the picture, or fill each section with a pattern.

OPTIONAL: Create a related math-art design of your own.

Malcolm Swan's Partition Products

Pick a number. Break it up into as many pieces (called *partitions*) as you want. Now multiply all those pieces together. Try a different set of partitions. What's the biggest product you can make?

For example, suppose your original number was 25. You might partition it into 1 + 1 + 1 + ... + 1, but that set has a rather disappointing product. (Why?) Or you could try 21 + 4, which multiply to make 84. Or 10 + 10 + 5, which makes a nice, big product of 500. Can you do better than that?

CHALLENGE: What if your partitions don't have to be whole numbers? For example, what if you split 25 into 2.25 + 4 + 6.25 + 12.5?

On This Day in Math

Go to pballew.blogspot.com and find today's "On This Day in Math" post. Choose one of the events or people mentioned and summarize that information here. Do a web search to see what else you can find out about that person or event.

Pixel Graphics

Outline a square that is 8 × 8 grid spaces or larger. Make a black-and-white or colored design by shading the grid squares. Each grid square represents one pixel, which must be totally filled or completely blank.

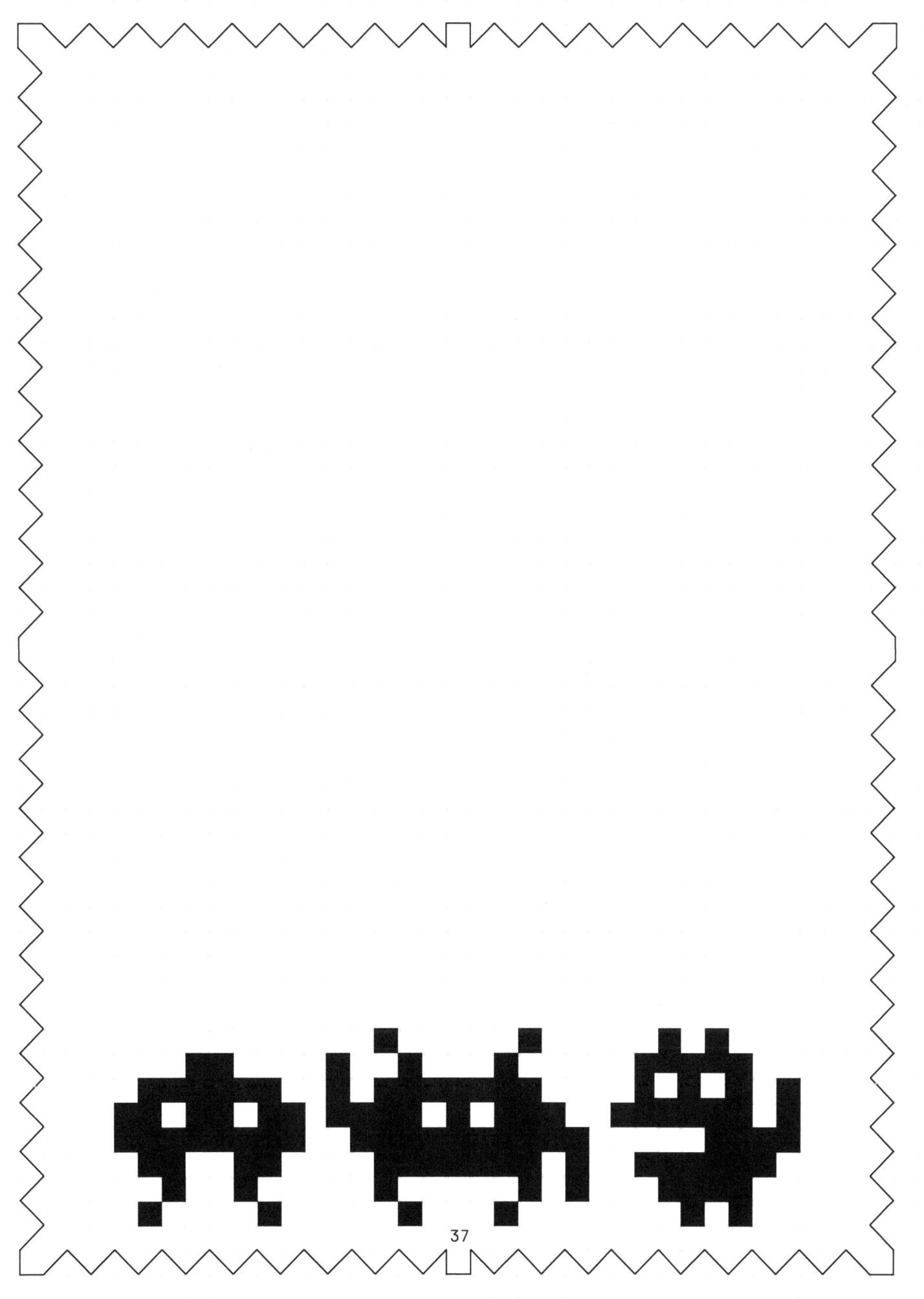

Explain a Problem

Copy a story problem from your math book, but don't include the numbers. Can you explain how to solve it without using any numbers?

For example: "Joseph knows the price of a box of candy and the price of a certain book. How can he figure out how much money he will have left after buying them both?"

One possible solution: "Joseph can add the prices together. If there's a sales tax, he also has to add in that percentage of the sum. This will give him the cost of his purchase. Then he can subtract this cost from the total amount of money he has, to find out how much will be left."

Extension: If you enjoy the challenge of solving problems without numbers, you might like Farrar Williams's book *Numberless Math Problems: A Modern Update of S.Y. Gillian's Classic Problems Without Figures.*

Numberstorming

It's like brainstorming about numbers. Write down everything you can about the number _____. You may include arithmetic expressions, expressions with words ("3 less than 10"), number properties (odd, prime, etc.), and other information ("days in a week," etc.).

Hieroglyphic Numbers

Look up how to write numbers with Egyptian hieroglyphic symbols. Write about how they work. Can you write your birth year Egyptian style? How would you do math calculations with numerals like these? Make an infographic using pictures and words to explain the hieroglyphic counting system.

Half Plus Three

Four children get pocket money. Each gets half as much as the next older child, plus $3 more. What questions can you ask?

Make up a fraction-plus-a-little-bit puzzle of your own.

Analogy

Finish the prompt sentence. And then keep writing until you run out of room. Don't overthink it, just write. Keep your pencil moving. If you can't think of what to write, copy your previous sentence over and over until your mind comes up with something new to say.

Math is like...

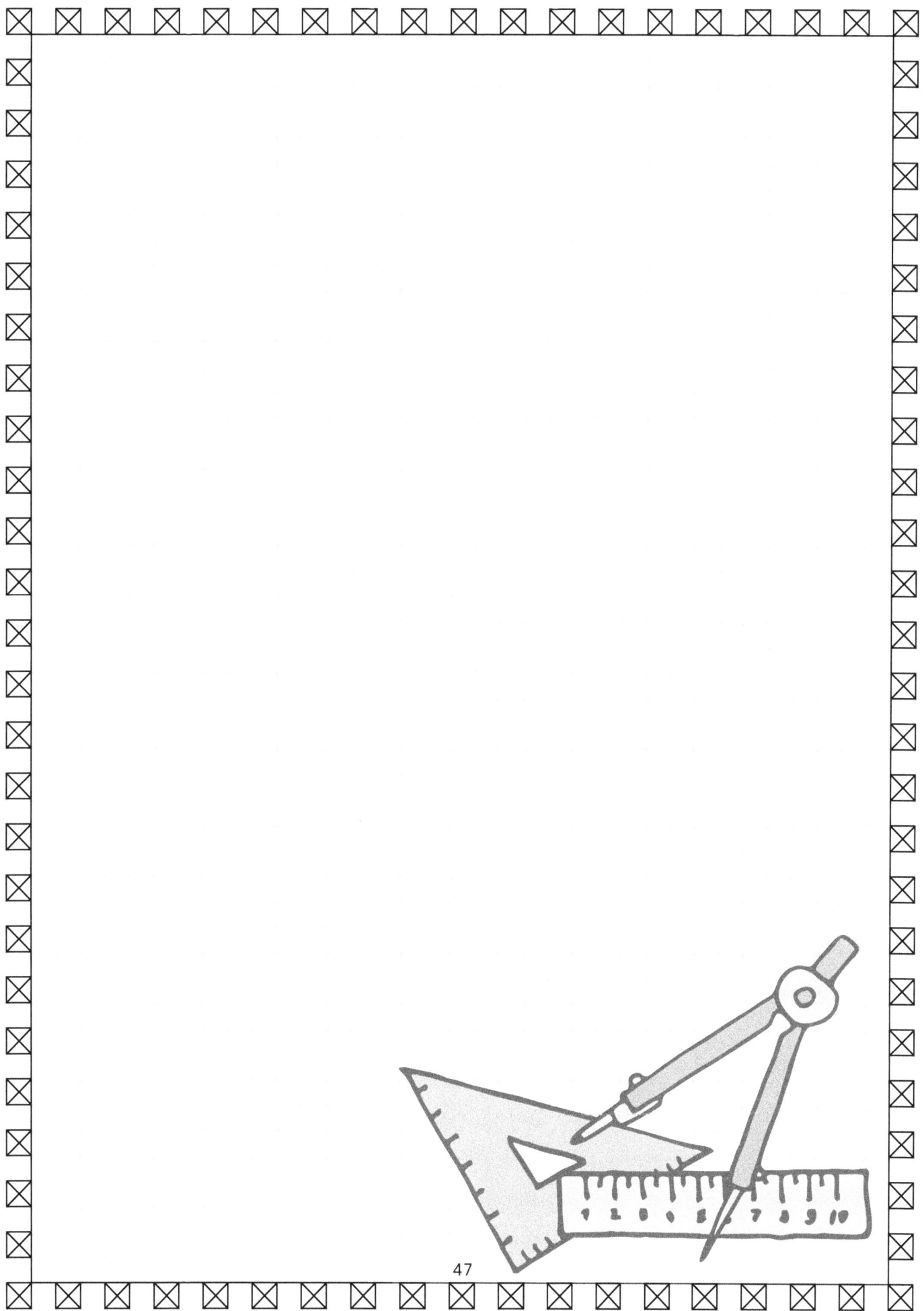

Sorting Collections

Collect a bunch of small items: buttons, Lego blocks, coins, etc. Dump them on the table. What do you notice about your collection? What categories or attributes might you use to sort the items?

Record your observations. What questions can you ask? How can you describe the collection with math? Would a chart or graph be useful?

Neutron Game

(two players)

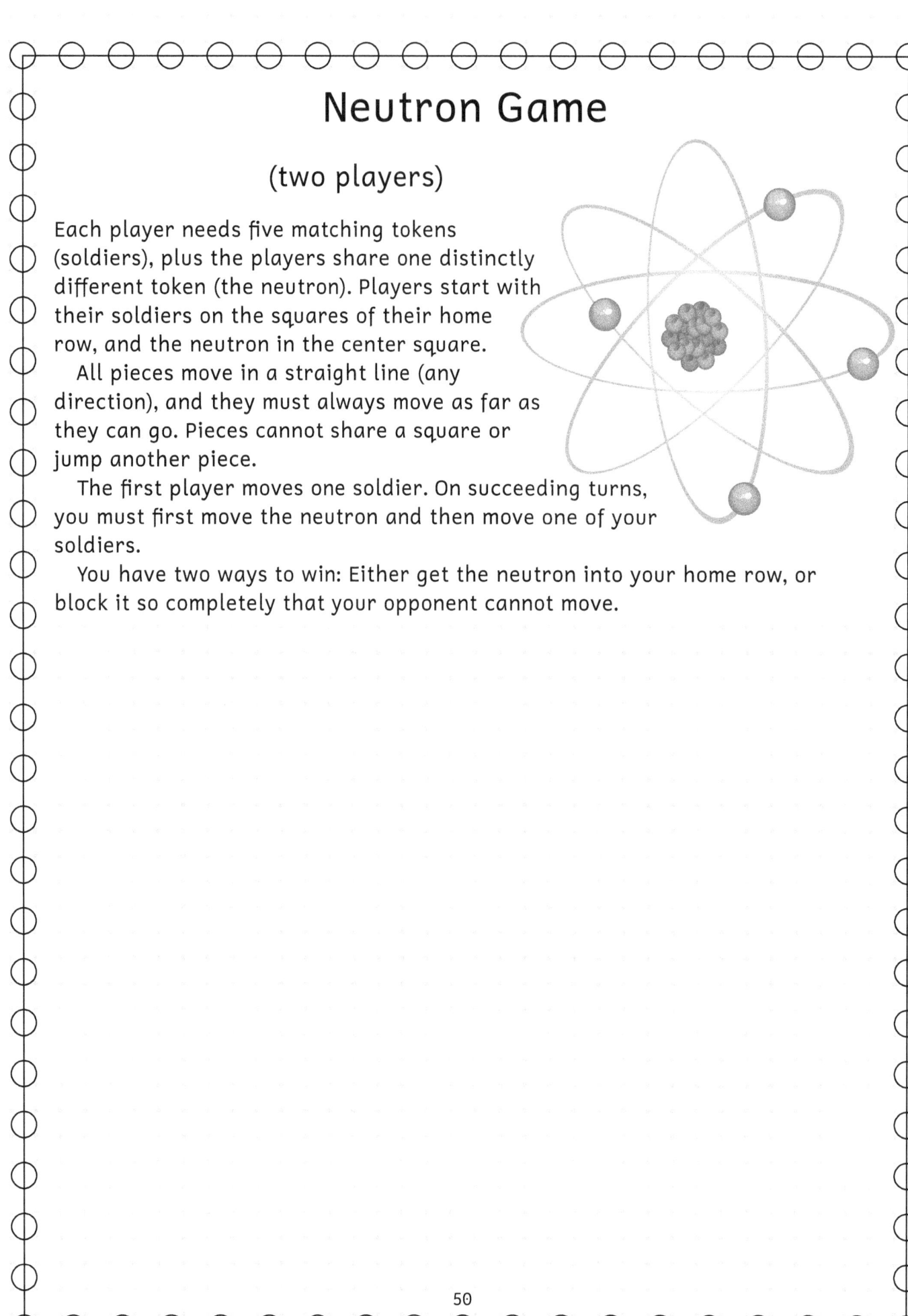

Each player needs five matching tokens (soldiers), plus the players share one distinctly different token (the neutron). Players start with their soldiers on the squares of their home row, and the neutron in the center square.

All pieces move in a straight line (any direction), and they must always move as far as they can go. Pieces cannot share a square or jump another piece.

The first player moves one soldier. On succeeding turns, you must first move the neutron and then move one of your soldiers.

You have two ways to win: Either get the neutron into your home row, or block it so completely that your opponent cannot move.

Player 1 Home

Player 2 Home

Area Puzzle

The rectangle has an area of _____ grid squares. [Choose any number.]
What size might it be? How many different rectangles can you find? What if the sides don't have to be whole unit lengths?

Julia Robinson Math Festival

Go to jrmf.org/puzzle. Choose a game or puzzle. Try it several times. Find different ways to play with the challenge. Write about your discoveries.

Candy Puzzle

You are working for a chocolate factory. You need to design a box to hold 48 candies. How will you arrange the box? Can you find more than one way?

Design boxes for different-sized candy assortments. What would be the hardest number of candies to package? Why?

My Secret Rules

(two or more players)

Choose a secret number rule for each of the circles in the Venn diagram. For example: even numbers, prime numbers, and multiples of 5. The other players take turns guessing a number, and you place it in the circle for the rule it matches.

Numbers that match more than one rule go in the section where the circles overlap. If the number doesn't match any of your rules, it goes outside all the circles.

Play until the others can name all of your secret rules.

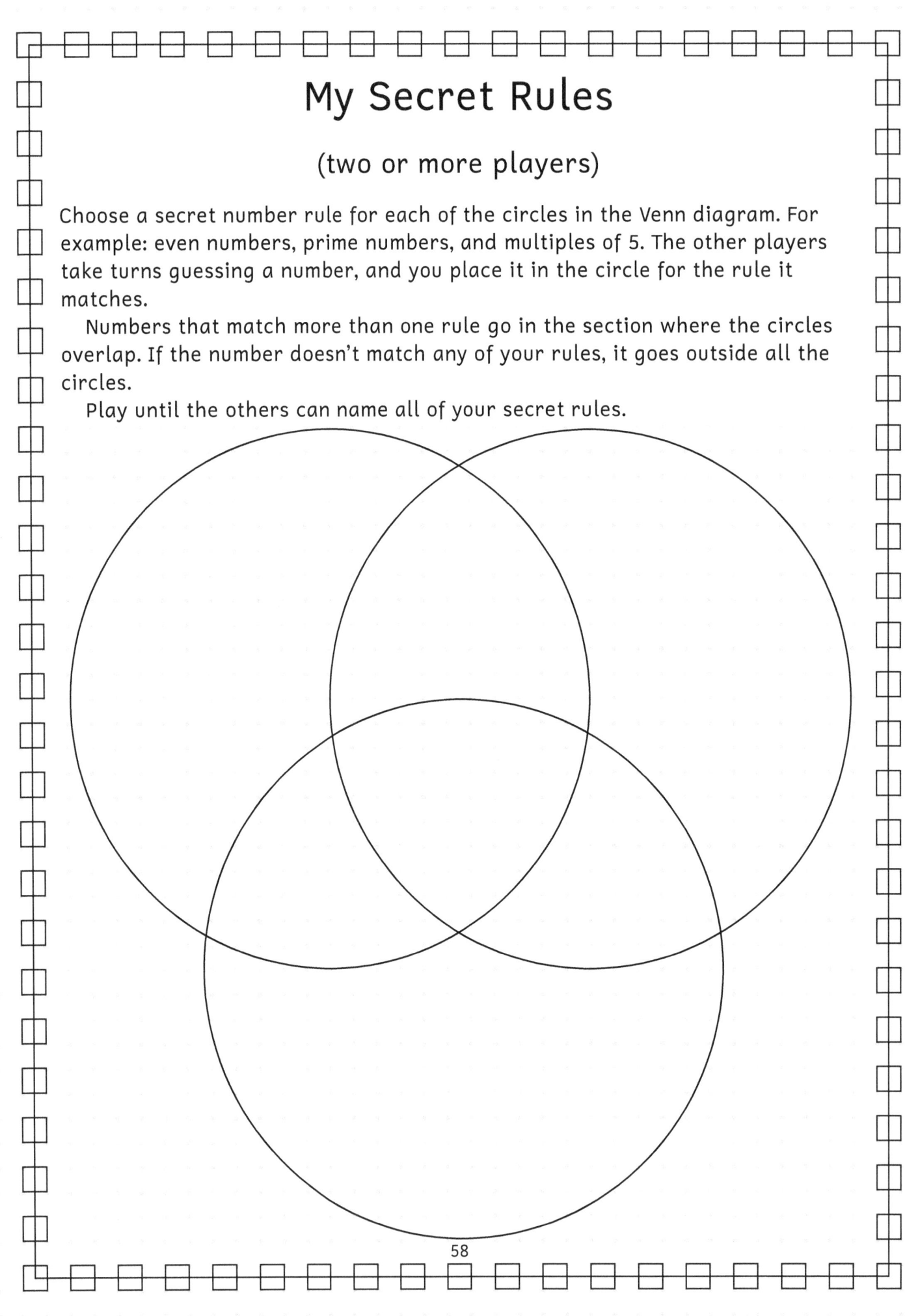

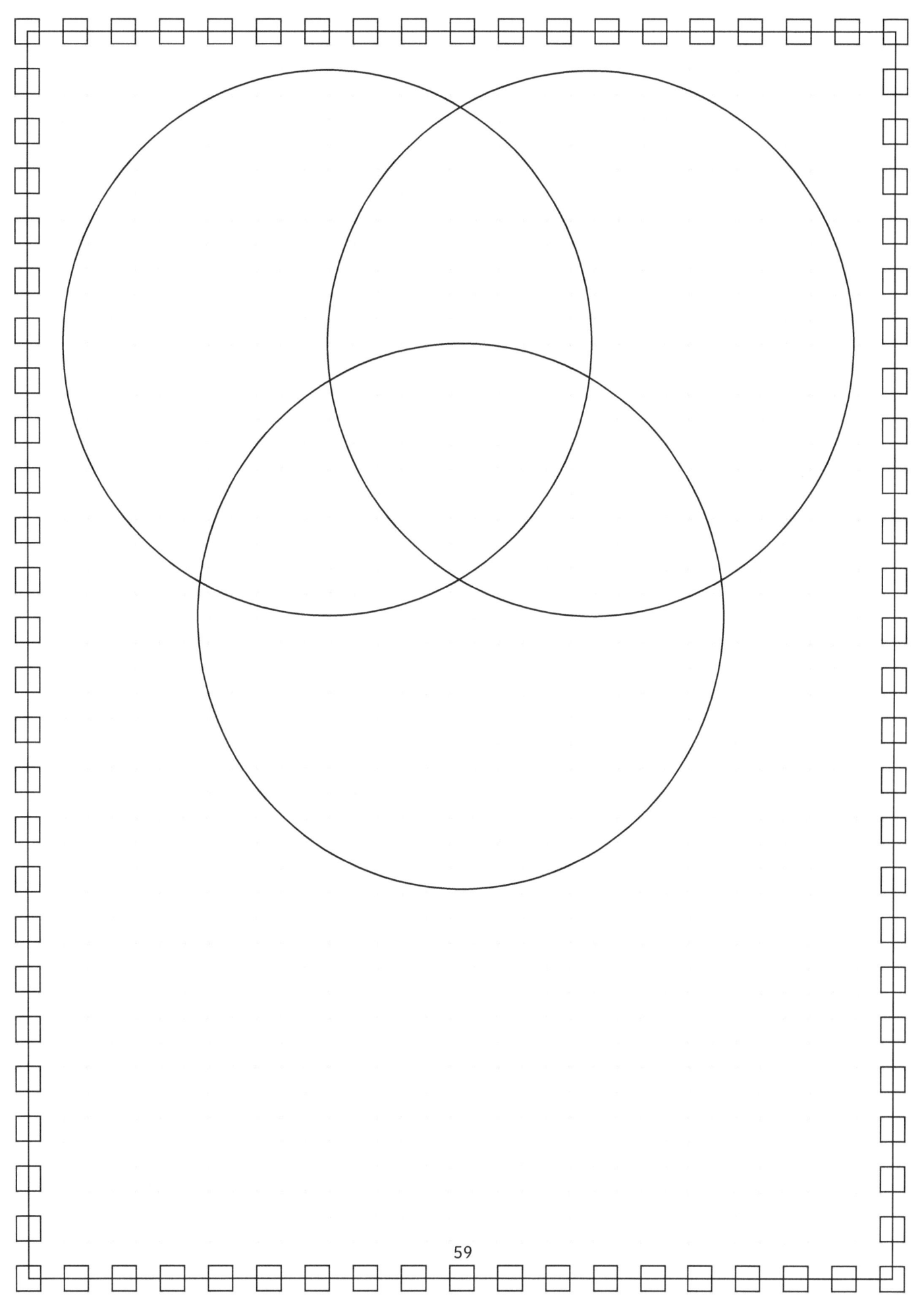

Number Yoga

Use the digits 2, 3, 4, and 5. You must have all four digits in each expression, with each digit appearing only once. You may use any math operation you know: +, −, ×, ÷, brackets, etc.

For example: 2 × 3 × 4 + 5 = 29, and 54 − 23 = 31.

Can you calculate all the numbers 1–10? 11–20? What other numbers can you make? Are any numbers impossible to make with your four digits? (Or perhaps they are possible, but not with the math you've learned so far?)

2-D Nim Game

(solitaire)

Draw a rectangle on your grid paper, between 10 and 20 squares in area. For example, a 3 × 5 grid makes a 15-square gameboard.

On your turn, cross out one or two of the grid squares—but you can only mark two squares if they share a side. Whoever gets the last square wins.

Misère variation: Whoever marks the last stone loses.

A Generous Gift

Pretend your grandparents gave you a penny on the first day of your birthday month, and two pennies on the second day, and kept doubling the pennies until your birthday. How big is that gift? What if they kept adding pennies for the entire month?

Fraction Pictures

Go to fractiontalks.com and find a puzzle you like. Make a sketch of it in your journal. Identify as many of the fractional parts as you can. Explain how you know the amounts.

Collecting Data

Choose something you are curious about that can be measured. For example: the outdoor temperature or rainfall, or how many jumping jacks you can do in a row, or how much time you spend on social media.

    Measure it every day for at least a week. Write a list of things you notice about the numbers that you find. Make a chart or graph to visualize your data. Remember to label your chart or graph, so people can tell exactly what you measured.

OLDER STUDENTS: Try several different styles of graph. Which one do you like best for this type of information?

Connect the Dots 1

Draw a picture by connecting grid dots on your journal page. Use as many or as few of the dots as you wish. What do you notice about your picture? Did you use symmetry? What shapes can you see?

Color as you wish, or fill each section with a pattern.

Take-Back-Toe Game

(two players)

You will need one six-sided die and 40 pennies, poker chips, or other small tokens. Place 10 tokens in each of the four squares in the middle column of the gameboard. The other two columns are the players' home squares.

On your turn, you may pass or play. To play, roll the die and move that many pennies from a single space to a vertically or horizontally adjacent space, except that you may not exactly undo your opponent's previous move. (Diagonal moves are not allowed.)

If any three squares in your home column contain the same (non-zero) number of pennies, you win.

Player 1
Home

Player 2
Home

Seeing Stars

Look at the math-art design. What can you say about the shapes or angles? Make a list of the things you notice. What do you wonder? Color the picture, or fill each section with a pattern.

OPTIONAL: Create a related math-art design of your own.

Fictional Math

Think of the characters in your favorite story. How would they use numbers, shapes, or patterns?

Would they cook, or go shopping? Might they build something? Would they decorate it with a design? What would they count or measure?

Make up some math problems about them.

How Crazy Can You Make It?

Choose any target number you like. Fill each shape with an expression that equals the target. Can you make some cool, creative math?

Math Eyes

Imagine you have special glasses that let you see everything through the lens of math. Choose an item from your room or something in your house. Don't tell what it is, but describe it using as much math as you can. Think about measurements, shape, position, designs or patterns, motion, etc.

CHALLENGE: Read your description to someone. Can they identify your object?

Pig Game

(solitaire or group)

Roll a die as many times as you want, adding the numbers to your score. Stop when you wish and pass the die to the next player. BUT if you roll a 1 before you stop, you lose all the points you added during that turn. The first player to reach 100 points wins.

EXTENSION: Mathematicians love to tweak games just to see what happens. How would you modify this game? Test out your new rules with a friend.

Stump an Adult

Write the hardest math problem you can think of. See if your parent or teacher can solve it.

Mini-Biography

Write about a female mathematician who is NOT Ada Lovelace. Look at mathigon.org/timeline or arbitrarilyclose.com/mathematician-project or mathshistory.st-andrews.ac.uk for ideas.

Aliquot Game

(two or more players)

The first player writes any whole number up to 100. The second player subtracts an aliquot part of that number, writing a new number underneath.

An *aliquot part* is a unit-fractional part of the number (such as one-half or one-fourth or one-seventeenth) which is itself a whole number. That means you can subtract any *proper divisor* of the number—any divisor less than the whole number itself.

Players continue to alternate, each subtracting an aliquot part of the current game number.

The player who is left without a move—because they face a number that has no aliquot part (that is, the number one)—loses the game.

Flipping Pancakes

You made a batch of pancakes, and each one turned out a different size from all the others. But they are all mixed in an untidy stack. How can you flip the pancakes to order them neatly from smallest to largest?

Pick any number of pancakes you wish, in any random order to begin with.

For example, this stack is arranged 2–1–4–6–3–5. But a perfect stack would be ordered 1–2–3–4–5–6.

One flip = Insert a spatula under any pancake and invert the selected cakes onto the top of the stack.

After the flip shown below, the pancakes are now ordered 4–1–2–6–3–5.

How many moves will you need to tidy up your pancake stack?

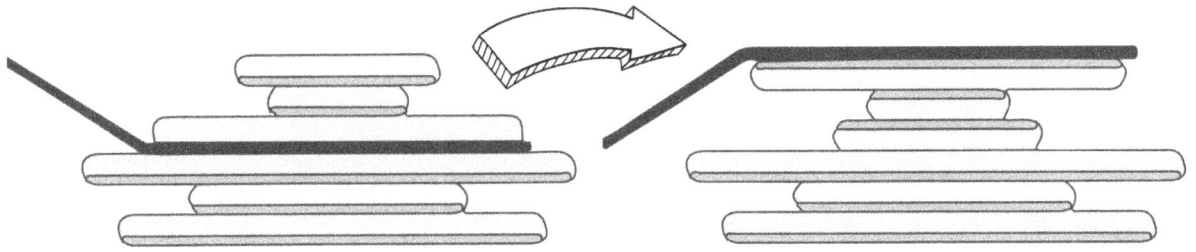

Reinvent Your Homework 1

Find a page of calculations in your math book, or download a worksheet online. Choose two or three of the questions. Write a story problem to match each calculation.

For example, for the calculation $3/4 \times 8$, you might imagine a recipe that takes $3/4$ cup of flour. But you are planning a party and need to make eight times that amount. How much flour will you need in all?

Sci-Fi Puzzle 1

Aliens from the planet Vargo have either 3 or 5 antenna-spikes. [Or choose different numbers.]

A Vargon spaceship with a diplomatic crew visited Earth. When they landed, news reporters counted the total number of spikes, but static interrupted their broadcast.

What might the number be? Which numbers are impossible?

Can You Solve It?

Go to expii.com/solve or search for the MathCounts Problem of the Week. Find a math problem you like. Copy it in your journal.

Or copy a problem from your math book.

Explain how to solve your problem. Try to make your explanation clear enough for a younger sibling, cousin, or friend to understand.

Challenge: Can you show more than one way to figure it out?

Growth Mindset

Have you heard that your brain keeps growing the more you use it? And that mistakes help you learn even more than when you get things right? How do these scientific discoveries affect your attitude towards math?

Quarter the Square

How many different ways can you divide a square into fourths? The four pieces must have the same area, but not necessarily the same shape.

Can you find a division with only one line of symmetry? With no lines of symmetry, but having rotational symmetry? With no symmetry at all?

Sort and classify your designs. What other questions can you ask?

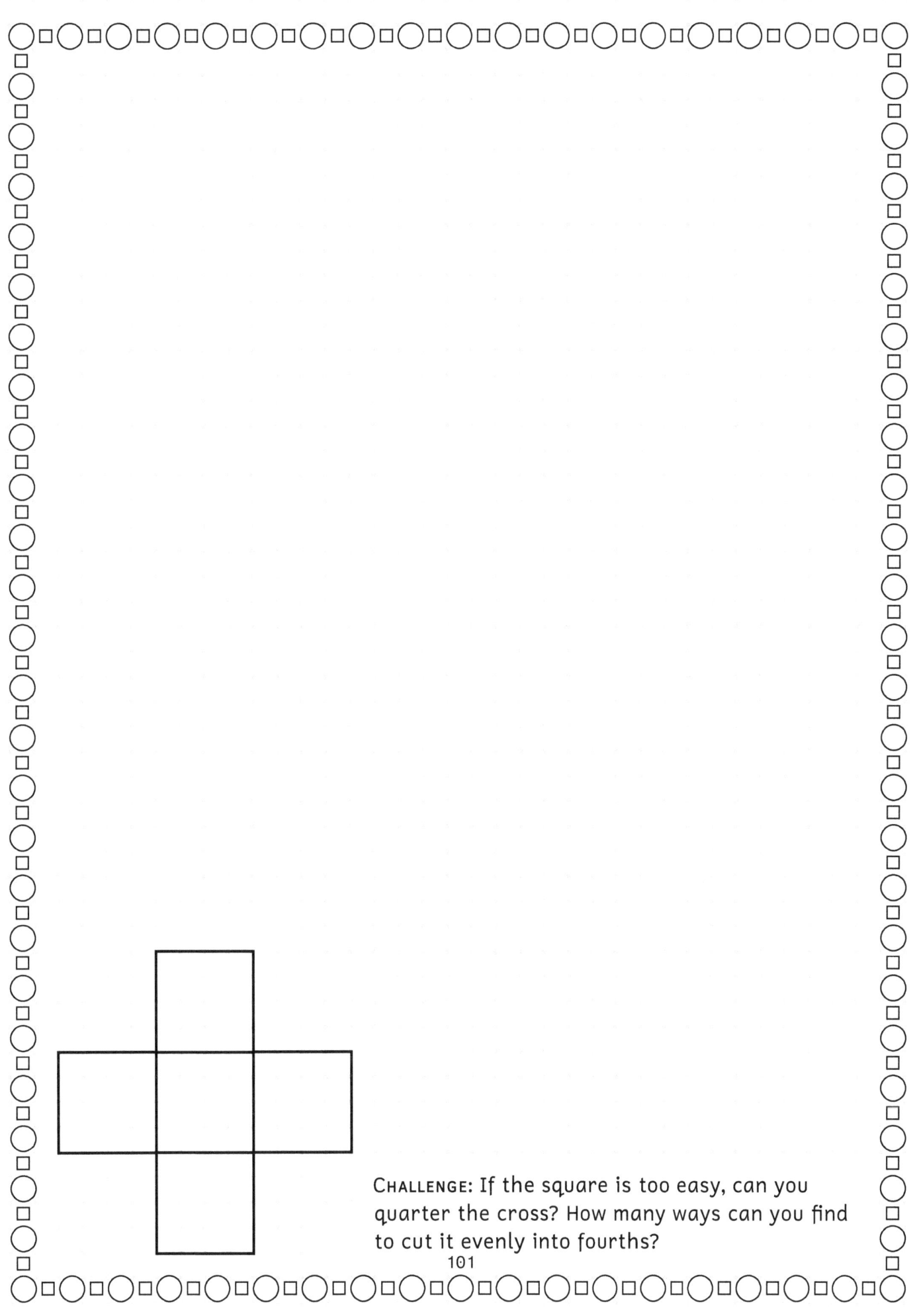

CHALLENGE: If the square is too easy, can you quarter the cross? How many ways can you find to cut it evenly into fourths?

Which One Doesn't Belong?

Go to talkingmathwithkids.com/wodb and find a puzzle you like. Make a sketch of it in your journal. Write out your answer(s) to the puzzle, using arrows or diagrams as needed.

Create Your Own WODB

Choose three attributes an object might have. For example, you might pick pointy, blue, and striped. Then draw four pictures. One picture should have all three attributes, and the other pictures should each be missing a different one. Ask a friend to say which one doesn't belong and why.

What Would You Choose?

Would you prefer a stack of quarters equal to your height, or a bag of quarters equal to your weight? Why?

Create some make-a-choice questions of your own.

107

Math Book Review

Find a math book at your library. Read it and write a summary or review.

If your library uses the Dewey Decimal System, mathematical nonfiction is on the 510–519 shelf, and logic puzzle books are at 793.74. You can ask the librarian to help you find a mathematical picture book or fiction story or the biography of a mathematician. Or do a search for "CSMP storybooks by grade level."

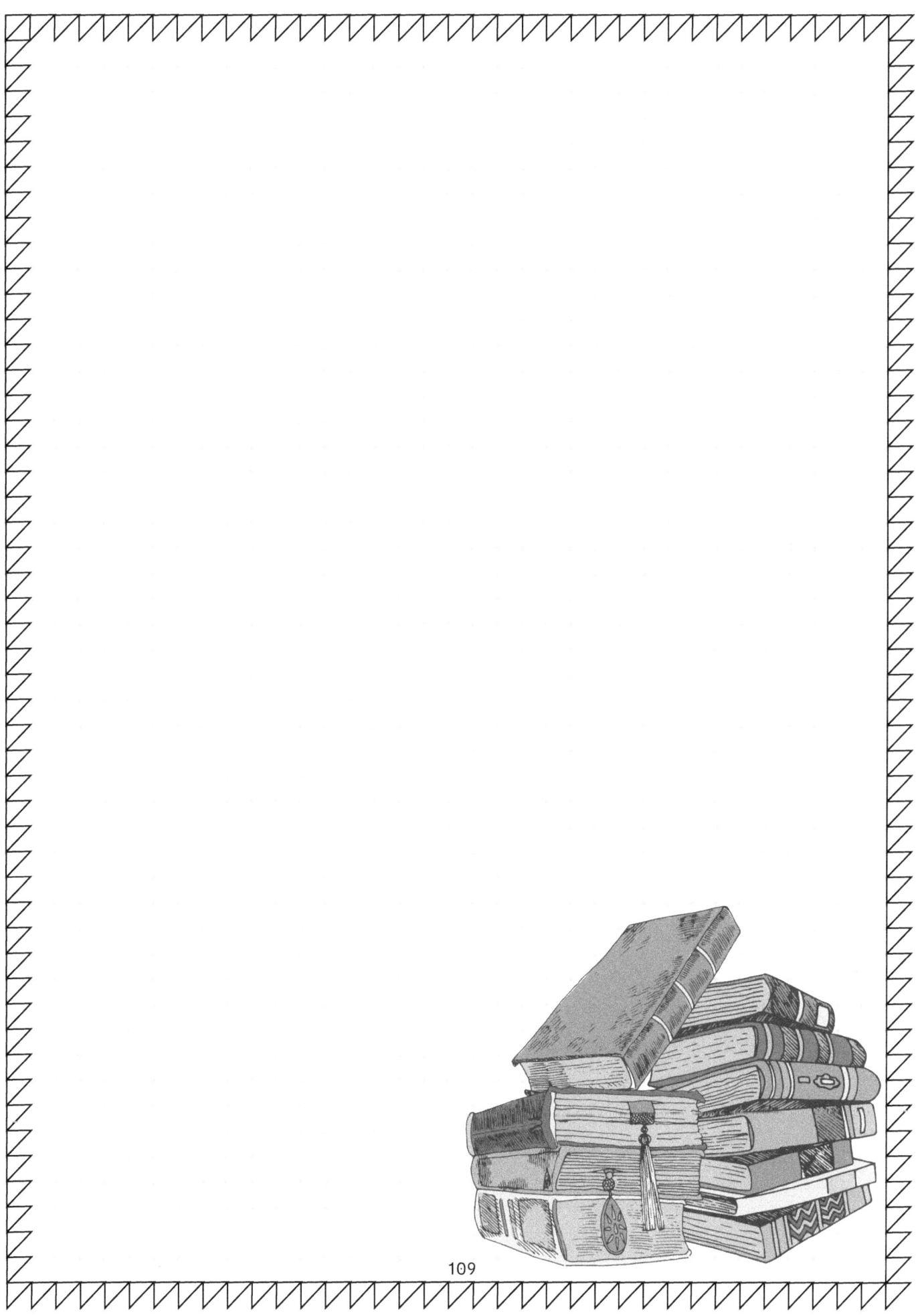

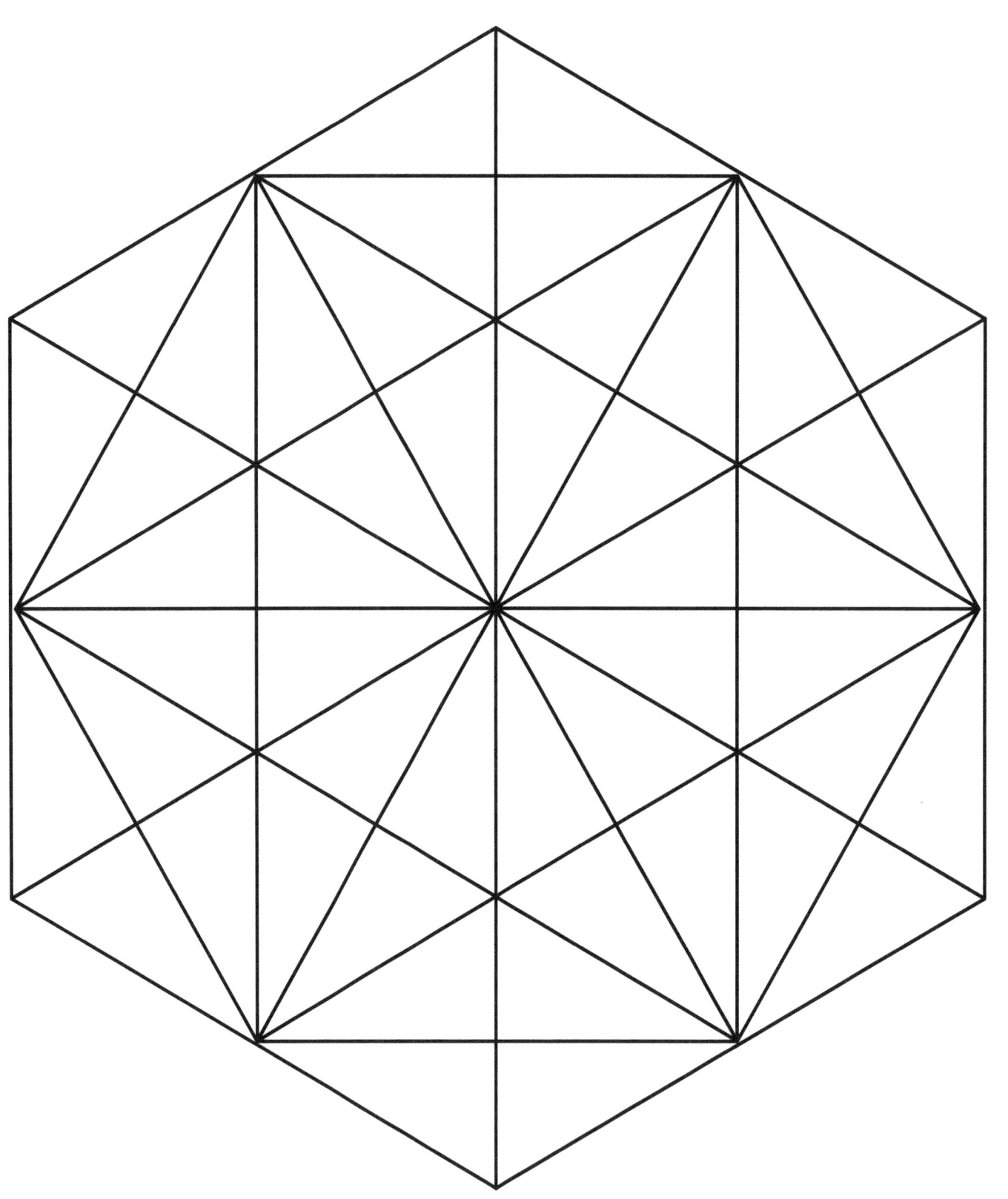

Hexashapes

How many shapes can you find? (Rectangle, rhombus, kite, etc.) How do the shapes relate to each other? For example, can you find different shapes that have the same area? Or a shape that is some fraction of another shape?

Contig Game

(two or more players)

On your turn, roll three dice. Use those numbers and basic arithmetic (+, −, ×, ÷) to calculate the number in any unmarked square on the board. Mark your square with a large X. Say out loud how you made the number.

Score 1 point for the square you marked, plus 1 point for each already marked square *contiguous* to your number's square—that is, touching any side or corner. The maximum score for any turn is 9 points. If all the numbers you can make have already been marked, you score a zero—but if anyone else finds a valid calculation using your dice, that player may mark the square, and "steal" those points.

When another player thinks you made a mistake, that person may challenge your answer before the next player rolls the dice. If your answer was wrong, the challenger takes the points you would have won, and you score zero. But if your calculation is correct, you get one bonus point for having withstood the challenge.

Play until each player has had ten turns, or five turns each for more than three players. Whoever has the highest total score wins the game.

Keep Score:

Contig Game Board

1	2	3	4	5	6	7	8
28	29	30	31	32	33	34	9
27	55	60	64	66	72	35	10
26	54	125	144	150	75	36	11
25	50	120	216	180	80	37	12
24	48	108	100	96	90	38	13
23	45	44	42	41	40	39	14
22	21	20	19	18	17	16	15

Explain a Puzzle

A stick has two ends. If you cut off one end, how many ends will the stick have left?

A square has four corners. If you cut off one corner, how many corners will the remaining figure have?

Is subtraction broken?

Sci-Fi Puzzle 2

Aliens from the planet Crimbat double their population every year. They send out lots of colony spaceships to keep from overpopulating their home.

A Crimbatti settlement on Earth contained 6,400 aliens in its 6th year. In which year was the population half that size? How many Crimbatti settlers were there in the first year?

How long until the aliens outnumber the humans on Earth?

Reiner Knizia's Banker

(two players)

Players each need paper, pencil, and a single die. Roll your die, and write that number down to begin your math expression. Roll a second time, and choose a calculation symbol (+, −, ×, ÷) and then write the new number.

Roll three more times, adding to your calculation with a different symbol each time. You must use all four math operations.

Calculate the value of your math expression. When dividing, ignore remainders. The player with the greatest value wins.

□ +−×÷ □ +−×÷ □ +−×÷ □ +−×÷ □

EXTENSION: Mathematicians love to tweak games just to see what happens. How would you modify this game? Test out your new rules with a friend.

Ratio Puzzles

The ratio of dogs to cats at the animal shelter is 2.4:1. How many cats and dogs might the shelter have?

What is another possible combination? Why can't you know the exact answer?

Make up your own ratio puzzles.

Connect the Dots 2

Draw a bunch of dots scattered around your page, numbering them as you go. Then connect the dots using an unusual rule.

For example, connect each number to its double: 1–2–4–8, then 3–6–12, and so on until you run out of numbers with doubles, then connect the remaining numbers in order.

Make up your own rule. Color your design, or fill each section with a pattern.

Make a Million

Draw six boxes in a row, each big enough to write a digit from 0 to 9. Below that, draw six more boxes, like this:

☐ ☐ ☐, ☐ ☐ ☐

☐ ☐ ☐, ☐ ☐ ☐

Draw cards or roll dice to get random numbers. Write each number into one of the boxes. (For 10s and face cards, write a zero.) Once a number is written, you can't move it. When all the boxes are full, add your two six-digit numbers.

How close did you get to making one million?

Compete with a friend to see who can get closest.

In the News

Read a news article. Write about how it connects to math. Are there parts of the story that could be counted, measured, or graphed? Are there data that indicate a trend?

Examples: slowrevealgraphs.com and nytimes.com/column/whats-going-on-in-this-graph.

Words Help Us Think

Make a math "word wall" on your journaling page. Write all the math vocabulary words you know. Decorate them with frames, or write them crossword-style or in different directions. Or make it fancy any way you like.

OPTIONAL: Include definitions or pictures explaining the hardest-to-remember words.

Mountain Ranges

A mountain range is defined as a row of one or more equilateral triangles with a total base of a certain length. For example, below are three mountain ranges of length 4.

How many other mountain ranges can you make that are 4 units long?

Try making other mountain range lengths. What do you notice? What questions can you ask about these patterns?

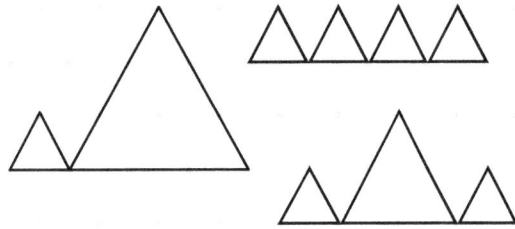

131

Shannon Switching Game

(two players)

Make a dot with the label "+" and a dot with the label "–", which are the two nodes of your game. Then put several dots spaced out in the area between those two. Draw lines connecting these dots however you like, so that every dot is connected to several others.

These lines represent the possible wires in an electric circuit.

One player tries to complete the circuit by connecting the two nodes. On each turn, strengthen one of the unmarked lines between two dots by drawing it thicker with your marker or pen.

The other player tries to break the circuit. On your turn, cut one of the unmarked lines by marking it with an "X."

The connecting player wins by completing a line between the nodes. The other player wins by making such a line impossible.

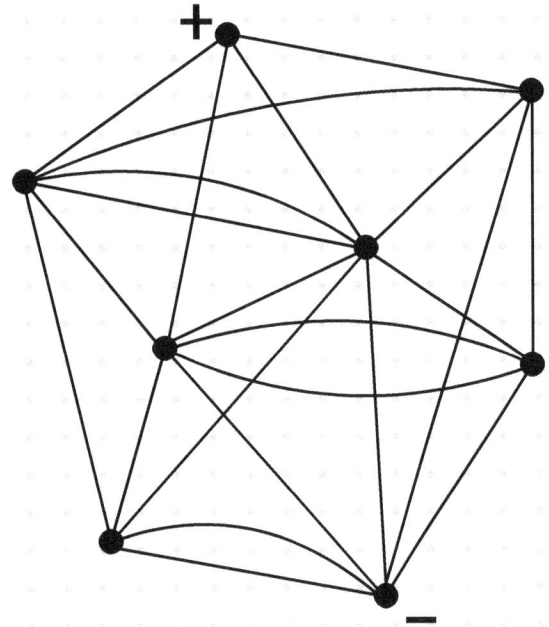

John Golden's Fraction Square

Outline a large square on your page. Draw lines to divide it into four or more parts, all different from each other. No two pieces can be the same size.

What fraction is each part, compared to the full square?

Do these fractions really add up to one complete whole?

Visual Patterns

Go to visualpatterns.org and find a puzzle you like. Make a sketch of it in your journal. Describe how you see the pattern growing. Use arrows or diagrams as needed.

   Pick a number from 50 to 100. Can you tell how many would be in that step of the pattern?

Create Your Own Pattern

Draw a pattern that grows according to some rule. Show the first three or four stages of your pattern's life. Can you describe the growing rule with math?

The Adventure of Learning

Finish the prompt sentences. And then keep writing until you run out of room. Don't overthink it, just write. Keep your pencil moving. If you can't think of what to write, copy your previous sentence over and over until your mind comes up with something new to say.

At first, I thought... But then I discovered... Now I wonder...

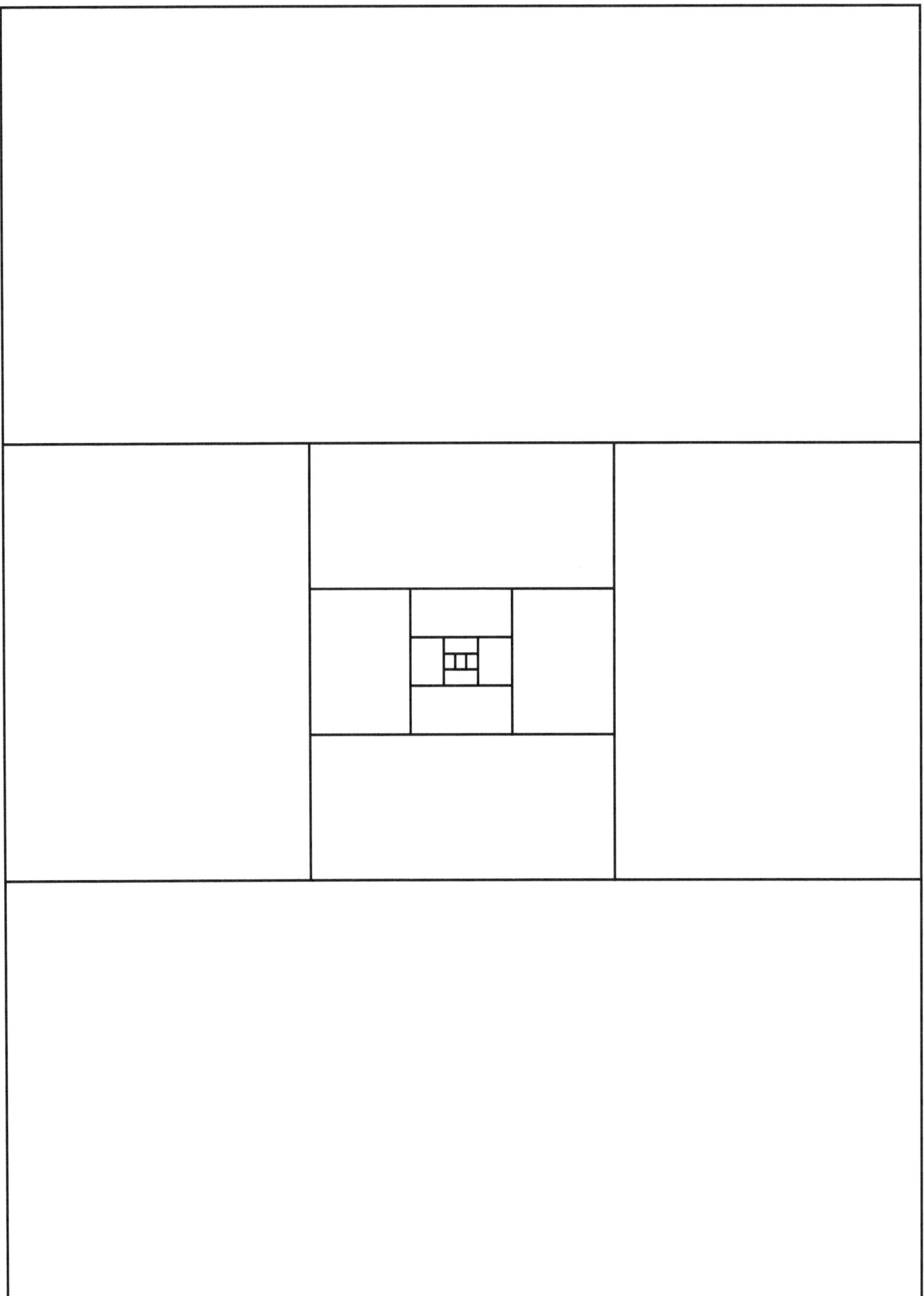

Infinite Series

Look at the math-art design. What can you say about the shapes or angles? Make a list of the things you notice. What do you wonder? Color the picture, or fill each section with a pattern.

OPTIONAL: Create a related math-art design of your own.

Lines on a Grid

With a ruler, draw a slanted straight line between two dots on your page. Can you draw another line parallel to the first? How do you know it's parallel?

Can you draw a line that's perpendicular? How can you be sure?

Make a design with parallel and perpendicular lines. Color as desired, or fill each section with a pattern.

Blockout Game

(two players)

Outline a rectangle of any size on your paper. Each player needs a different colored pencil or pen. You will also need two six-sided dice.

On your turn, roll the dice and multiply them. Color one completely connected shape with that area on the rectangular gameboard.

If your shape won't fit, you miss that turn. The game ends when there are two missed turns in a row.

Whoever colored the greatest total area wins.

EXTENSION: Mathematicians love to tweak games just to see what happens. How would you modify this game? Test out your new rules with a friend.

Museum of Math

Did you know there is a National Museum of Mathematics in New York? You can learn more about it at the website momath.org.

Imagine you are the curator designing a new exhibit for the Museum of Math. What will you put in your display? Will you make it interactive? How?

Exponential Halves

When you tear a piece of paper in half, you get two pieces. If you put one piece on top of the other and tear the stack in half, then you'll have four pieces.

If you were as strong as Superman, you could keep tearing the stack as it gets taller and thicker. How many times would you have to tear the paper to get 64 pieces?

What other questions can you ask?

Make It Visual

Draw a picture or diagram to explain something you've learned in math. Label it so anyone could look at your picture and know what you mean.

Hexangles

How many angles can you identify without using a protractor? Measure the angles to see if you were right. Measure some that you weren't sure about, to see what size they are.

How do the angles relate to each other? For example, can you find different angles that have the same size? Or an angle that is some fraction of another angle?

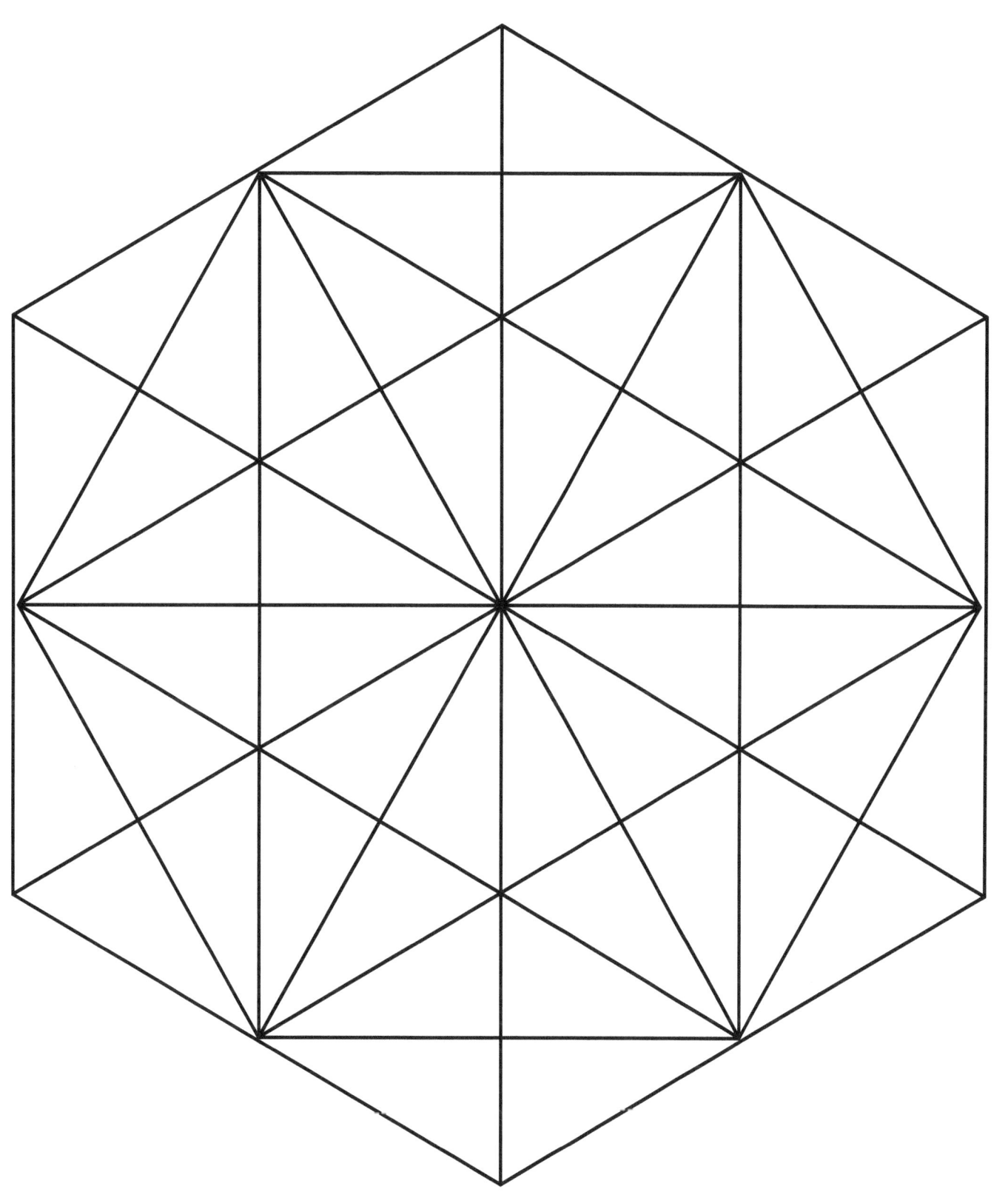

US Presidents

Who do you think lived longer, presidents from long ago or from more recent times? [Not from the US? Use the rulers of your own country.]

Now check the data, to see if your prediction was correct. Make a list of all the dead presidents and the age at which they died, then split them into two groups for comparison. What do you see?

HINT: A stem-and-leaf plot or box plot of the data may help when comparing the older and modern groups.

Math Art Challenge

Go to arbitrarilyclose.com/home or do a search for the #MathArtChallenge. Find a math art project you want to try. What do you notice? What do you wonder?

That's Mean

The mean (average) of four numbers is _____. [Choose any value.] When you add another number to the data set, the mean goes up [or down] by _____. [Choose a smallish value: 1, 2, 5, etc.] What was the sixth number?

Explain How

Think of something you have learned to do in math. Write an explanation simple enough that a younger sibling, cousin, or friend could understand the idea. Add pictures, charts, or diagrams if they help to make your meaning clear.

Mental Math Workout

Practice your Math Rebel skills. Pick any number, and see how many different ways you can write it. Start with a simple expression like addition or multiplication. Write an equal sign and another expression. Think about ways you can modify each expression to create a new one, and keep going until you fill the page or run out of ideas. What kind of fancy math will you create?

☐ =

Secret Number Codes

(two players or two teams)

Each player or team chooses four secret numbers less than 20.

A = _____, B = _____, C = _____, D = _____.

Take turns asking for algebra clues like "What is A + B + C?" or "What is D × A?"

But don't ask "What is A + 3?" or "What is ½ of C?" Players may decline to answer any question that directly gives away one of their numbers.

The first to guess the other player's code wins. Or just play until you solve both codes.

Same but Different

Go to samebutdifferentmath.com and choose a topic you've studied. Find a puzzle you like and make a sketch of it in your journal. Compare the two parts. Explain how the images or expressions are the same, and also how they are different from each other. How many things can you notice?

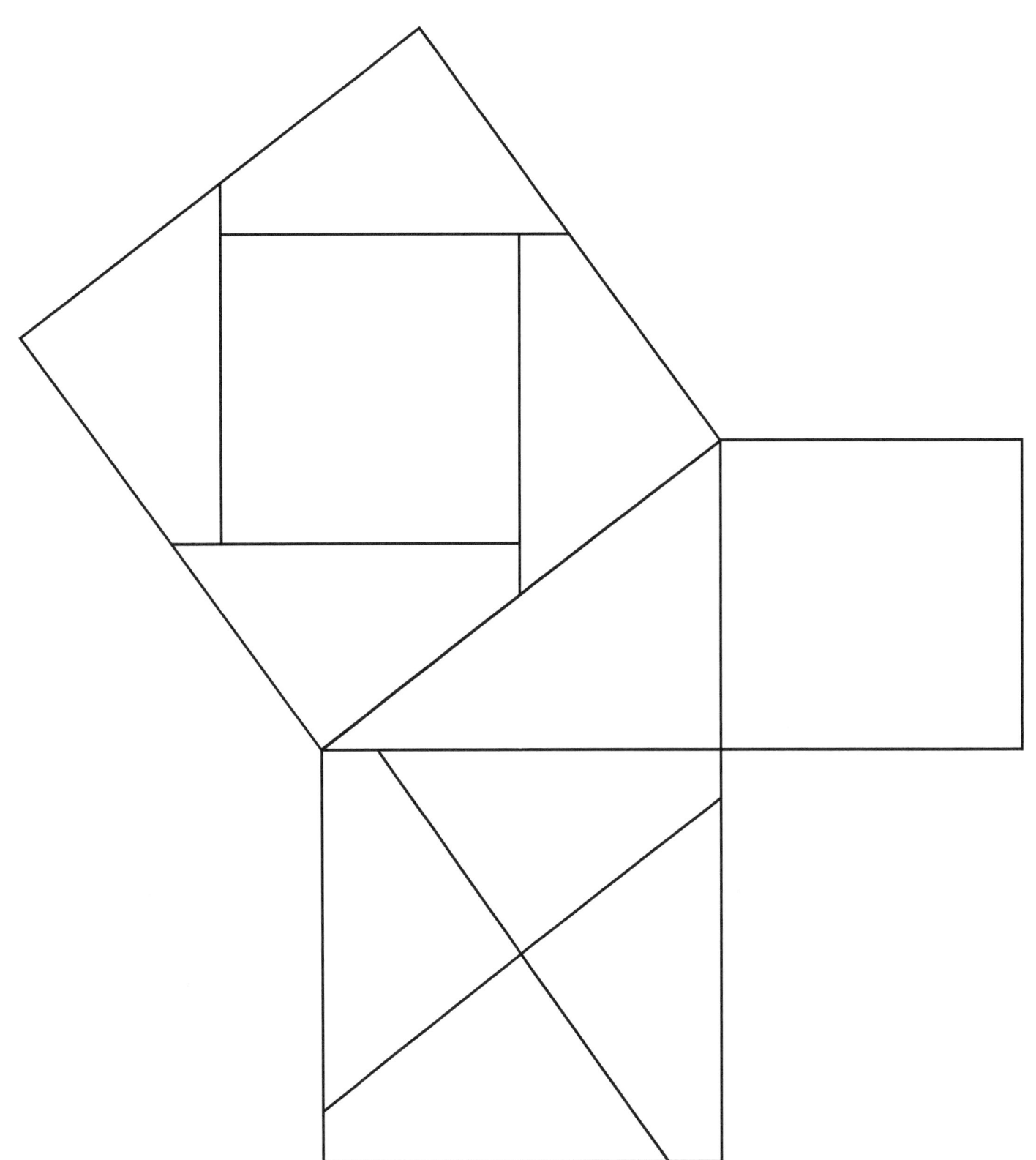

Cut and Paste

Look at the math-art design. What can you say about the shapes or angles?
Make a list of the things you notice. What do you wonder?
   Color the picture, or fill each section with a pattern.
   [Diagram by Henry Perigal (1801–1898).]

OPTIONAL: Create a related math-art design of your own.

Exponential Folds

If you fold a piece of paper in half, the result is twice as thick as the sheet by itself. If you fold it in half again, it will be four times as thick. Another fold makes the entire stack eight times as thick as a single sheet.

If you started with a huge sheet of paper, and if you were as strong as Superman and could keep folding forever, how many times would you have to fold the paper in half for the stack to reach the moon?

Math Poetry: Haiku or Senryū

Count syllables to make a non-rhyming mathematical poem. Haiku have three phrases (lines) with seventeen syllables in a five-seven-five pattern. Usually, the theme of the poem is expressed by a key word at the end of one phrase.

Senryū is haiku's more cynical cousin, exposing human foibles and often laced with dark humor. For example:

Multiplication

How much ice cream in

Five triple-dip fudge sundaes?

Getting fat with math.

All Twos

Use only the digit 2, and try to use as few of them as you can for each calculation. You may use any math operations you know. Can you calculate all the numbers from 1–20? What other numbers can you make?

Math Pickle

Go to mathpickle.com and click the link for "Puzzles," then choose your grade level—or just browse the website. Find a puzzle you like. Try it several times. Write about your experience, what you noticed and what you wondered about.

The Power of a Pattern

Choose any base number and investigate its powers. For example, the powers of 3 are $3^1 = 3$, $3^2 = 9$, $3^3 = 27$, $3^4 = 81$, etc. Extend the list as far as you can. What do you notice? What questions can you ask?

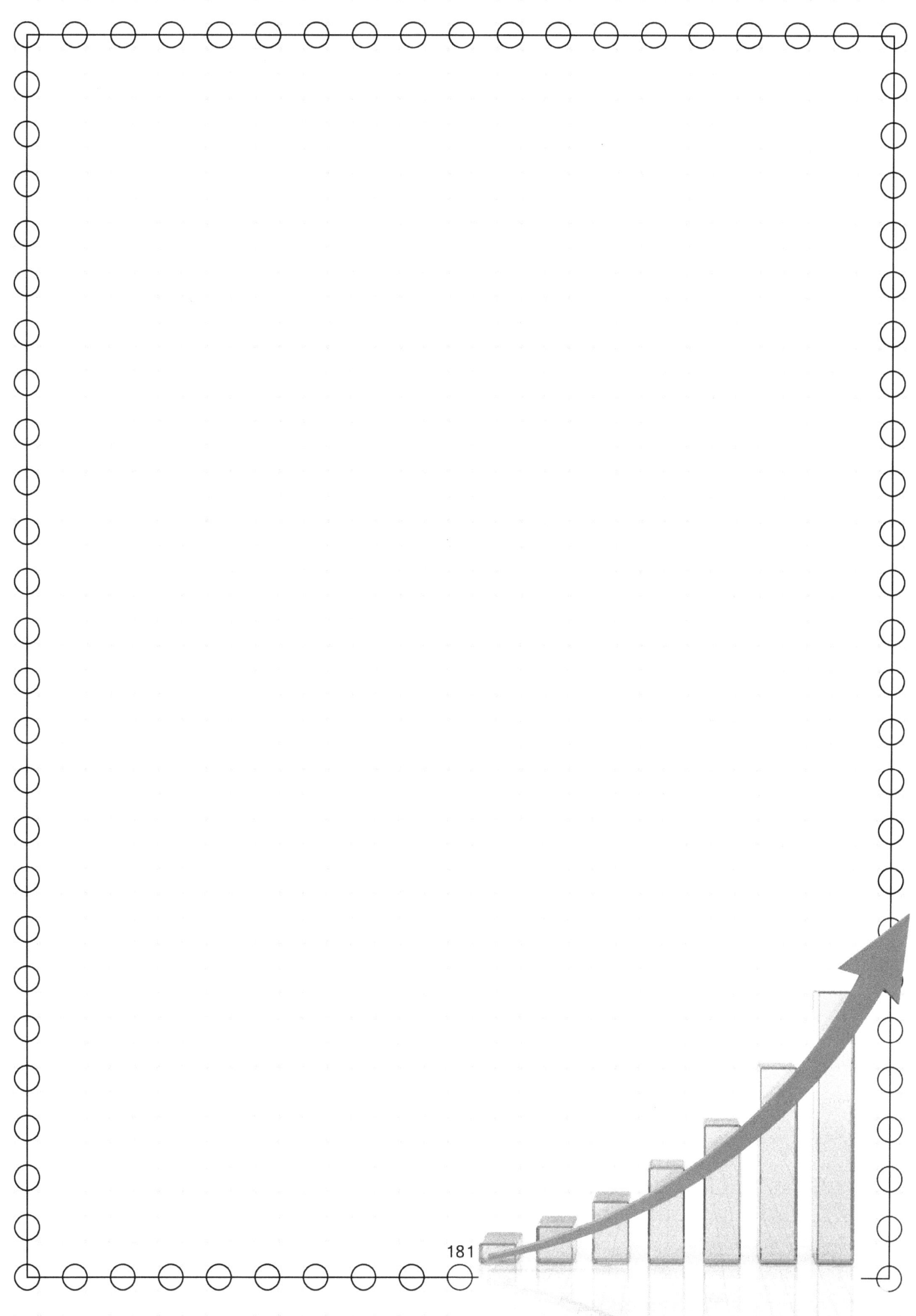

181

Brain Dump

Tell what you know about _____. Or, what are the most important things to understand about _____? [Choose any math topic, especially something broad like "addition" or "percents."]

Octagons

Connect dots in a stop-sign shape. Are all the sides of your octagon the same length? How can you tell?

Or make any 8-sided figure you like. Is your octagon convex, or did you make it concave? How many different octagons can you make?

Will your octagon tessellate? That is, could you cover a floor with only octagon tiles? How do you know?

What kind of design can you make with octagons?

Math Report

Read a math article. Look up something on Mathigon.org or Nrich.maths.org or MathsIsFun.com. Or search for the old Math Munch blog or Martin Gardner's classic Scientific American columns. Write about what you learned. What questions can you ask?

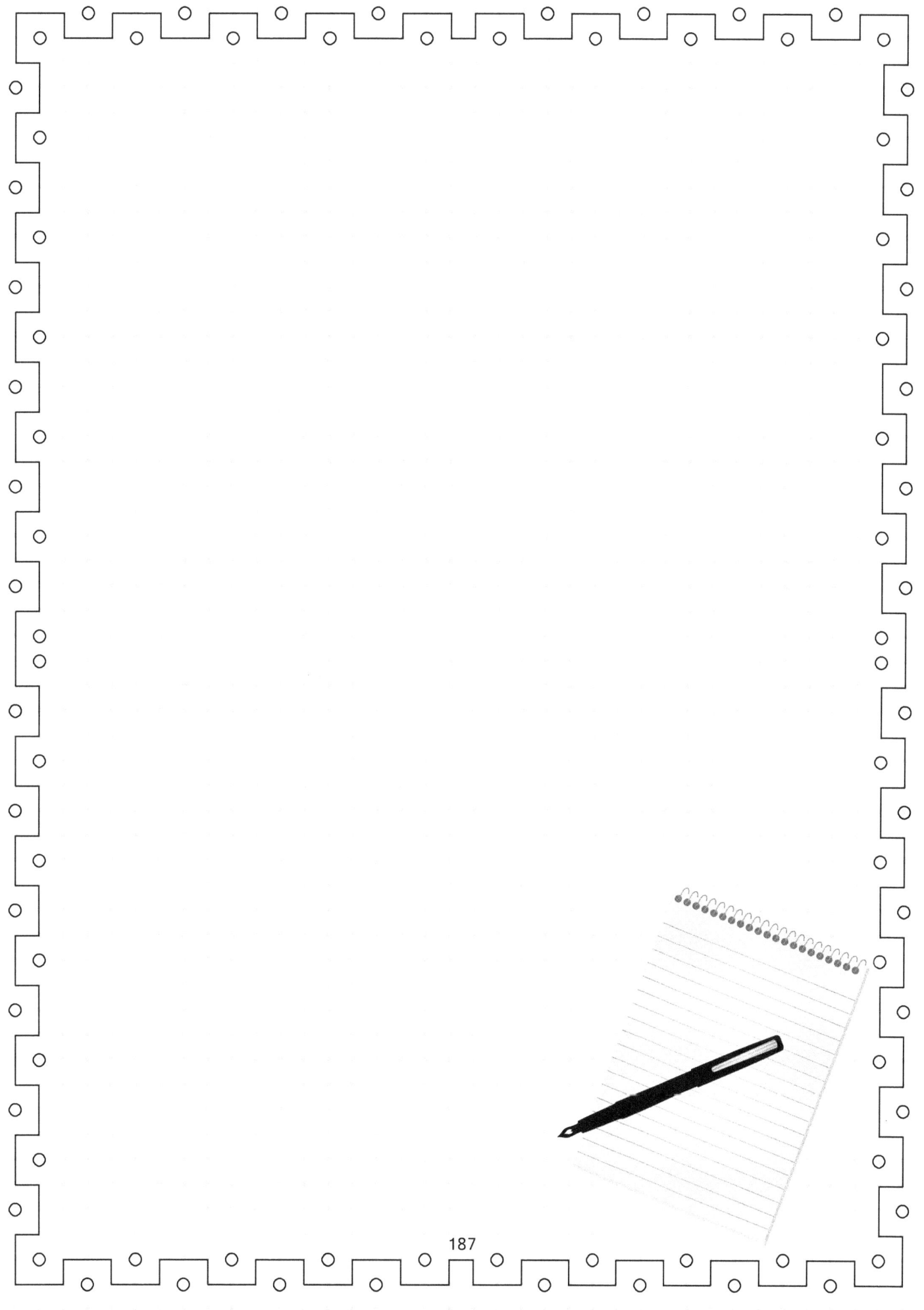

Frayer Model

Choose a math vocabulary word. Define it with a Frayer Model. Write the word in the middle of your page and circle it. Then draw lines to divide the remaining page into quarters with labels: definition, characteristics (or illustration), examples, and non-examples. Fill in each section.

Which section is easiest to do? Which takes the most thought? Can you think of anything else someone may need to know in order to fully understand the concept?

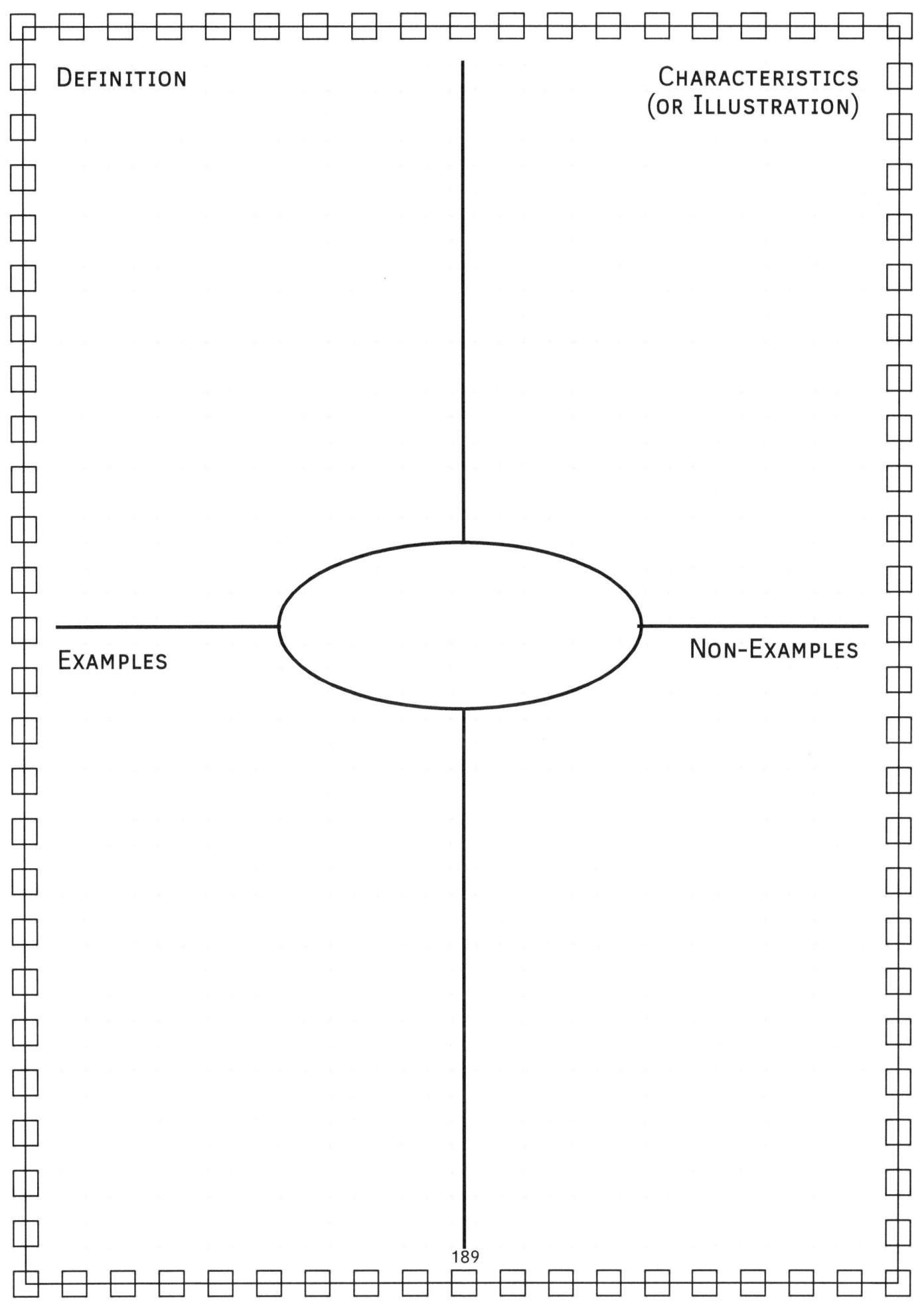

Monthly Math

Create a math calendar with a puzzle for each day of the month. Can you make each answer equal the number of that day?

Monday	Tuesday	Wednesday	Thursday	Friday	Saturday	Sunday

Place Value Mastermind

(two or more players)

Choose a 3–6 digit secret number. Write your number down, and tell the other players how many digits it has. Each player in turn will guess a number with that many digits.

Write the guess below your secret number, and then tell the others how many digits are exactly correct (but not which ones!) and how many are almost correct (the digit is in your number but not in the right place).

Whoever correctly names your secret number gets to be the next Mastermind, choosing a number for you (and the others) to guess.

Guesses	Number of Correct Digits	How Many Almost Correct
My Number:		

Place Value Mastermind

Keep track of the clues when you are trying to guess another player's secret number:

Guesses	Number of Correct Digits	How Many Almost Correct

Math Riddles Redux

(any number of players)

Choose a secret number the other players will try to guess. Write a "What Number Am I?" riddle. Give at least three clues for your mystery number. No other number should match all the clues.

?

Debate with Sofya Kovalevskaya

Decide whether you agree or disagree with the quotation below—and then argue the other side. If you agree, explain why someone might disagree. Or if you think the quote is wrong, tell how someone might argue that it's true.
   [You may also present your own point of view, if you like.]

> It is impossible to be a mathematician without being a poet in soul.
>
> —Sofya Kovalevskaya

Reinvent Your Homework 2

Find a page of calculations in your math book or download a worksheet online. Answer each question Math Rebel style: Write any true statement *except* what the answer key expects. Have fun making crazy math.

Captain's Log

Keep a math log for a week. Write down all the math you do each day, not just schoolwork. Try to write at least one sentence every day.

Did you learn anything new? Did anything surprise you?

Classify the Aliens

While exploring the alien world of Glorf, you meet a variety of local beings, many of whom move by floating in the heavy atmosphere. Your job is to classify the species of Glorfian creatures. How will you sort and organize them?

Urbanization Game

(two players)

Draw a box frame around a grid of 6 × 6 dots, the home ground of your new city. Take turns to connect two adjacent dots (diagonal lines are allowed in this game), or a straight line of three adjacent dots.

If you draw the final line that encloses an area, claim that property by marking it with your symbol or color. The outer border does NOT count as part of the game area, but only identifies the boundary of play.

Play until the grid is full. The player whose claimed property sections have the largest total area is the winner.

VARIATION 1: In the traditional game Dots and Boxes, any player who claims a box must take another turn. Try playing Urbanization with that rule. How does it change the game?

VARIATION 2: What if the line of three dots is allowed to bend?

Strategic Thinking

Play any math or logic game. Describe your strategy for winning. How do you plan your moves? How do you figure out responses to your opponent's moves?

Don Steward's Rooftops

Rooftops are made from square blocks. Four examples are shown below. What do you notice? What questions can you ask?

How many different rooftops could you build with 24 blocks? [Or pick another number, or some other constraint.]

Is there a mathematical rule for rooftops? What measures might vary? What always stays the same?

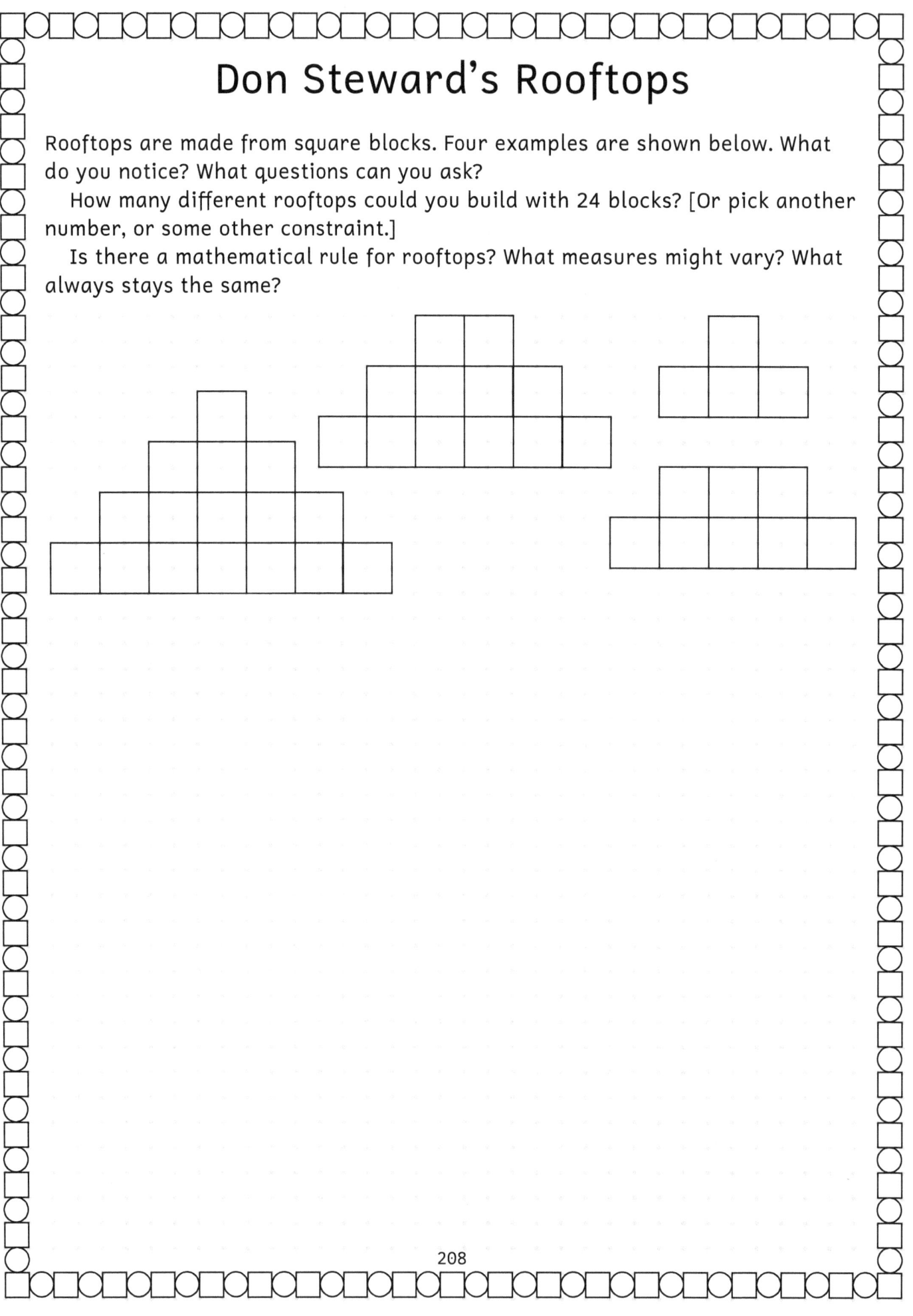

The Substitution Game: Round 2

(solitaire or small group)

Begin with a complexified equation from an earlier round of the Substitution Game. On your turn, copy the equation from the line above, except replace any short expression with an equivalent single number. For example:

$$2 + 8 - 3 = (10^2 - 86) \div 2$$

$$2 + 8 - 3 = (100 - 86) \div 2$$

$$10 - 3 = (100 - 86) \div 2$$

If you end with a true expression like "2 = 2," you win. Congratulations!

If you end with a false expression like "2 = 7," then at least one person made a mistake somewhere in either round of the game.

Don't worry about trying to find the mistake. You still win, because you got to play with a lot of wild and crazy math. So accept the reminder that you're human, and enjoy a laugh at your silly result.

Rear-View Mirror

Has your mathematical imagination grown as you explored the activities and questions in this journal? Explain how, or discuss why not.

If you were the next Math Team leader, how would you guide your students on a creative mathematical adventure?

Special Thanks

This book came to production with the help of many wonderful people who backed the project on Kickstarter. Your generous support and encouragement keeps me going!

I'm especially grateful to:

Abigail, Eleanor, and George Thompson
Ada Brinker
Adele & Davide
Adeline & Ethan Guilford
Adnane
Alexandra Stevens
Alina and Alice
Angela Z.
Annabelle, Mary Grace, & Evelyn Stephens
Ariana Anderson
Bair Family
Biswas Family
Debbie, Charli & Maddox Fear
Donnell Family
Edgewood Academy
Elizabeth Do
Elizabeth N.
Emelie Thomas
Gilles Durand
Jo O
Jolie, Dimitri, & Dominic
Jonathan, Noah, Faith, Joy, Jeremiah, Nathaniel, Grace, and Hope
Josiah, Alycia, Makelle, Jaxon, Eden, & Asher
Kayleigh & Aidan Cash
Lanie Toth
Lily, Juliana, Arizona, Peyton, Seth, Harper, & Hannah
Lindy Rose Capel
Lucas Dietrich
Lynne Menechella
Malcolm and Calvin Faull
MatematicasVedicas.org
Matilda & Willow
Matthew, Elizabeth, Mark, Evelyn, and Morgan Johnson
Moreno Family
Moriah, Tobias, Abram and Rosalind
Natural Math Alliance
Pattie Perry
Reynolds Family
Stroman Family
The Abraham Clan
The Bancks Family
The Beaumonts
The Berman Family
The Chesebro Family
The Childers family
The Cook Family
The Mickle Family
The Moore Family
The Powell Family
The Prentice Pack
The Santos Family
The Whitsell Family
Thomas Family
Tracy Popey
V.G.K.
Villarreal Family
Wells Family
Zëiss, Tesla, Hans, & Luna Knowlton

thank you grazie gracias merci efharisto spasiba danke cheers arigato
tack blagodaram marahaba nkosi shukria merkzi köszönöm kiitos takk manana sulpäy danki nuhun waybale gratzias shakkran tanmirt barkal ahsante nuhun dankewol bedankt
dankegon yekeniele dziekuje shukriyaa dziakuju stuutiyi matondi modupe menlau sobodi ngiyabongashukran matondo buznyg grassie obrigado miigwech tanmirt manjuthe chokrane dakujem maururu
madlobt misaotra spas mutumesc nandri blagodaria talofa vinaka aabar
waita obrigada spas dèkoju gràcies welalin zikomo ahsante tenki akiba
murakoze soolong kinisou rahmat wado supas skee sadol
mahalo salamat tänan saha dankewol tashakor
dankie mèsi
aitäh taiku
omol dèkuji paldies

Discover the World of Mathematics

Textbooks make math feel like a ladder to climb, working rung by rung from one topic to the next. But that's an illusion. Learning math is more like taking a meandering nature walk. Students need to wander around the concepts, notice things, wonder about them, and enjoy the journey.

Most people believe the goal of math is to get right answers. So teachers give us a page of math problems to work, then they check our answers and tell us what we got wrong. When math focuses on right answers, it can make anyone feel like a failure.

Instead, math lessons need to focus on reasoning, on how we struggle through the problems and figure things out. And when mathematical thinking is our goal, those right answers will come along as a side-effect, the natural result of making sense of the math.

—Denise Gaskins

Denise Gaskins' Playful Math

Tabletop Academy Press publishes playful math books for parents who want to help their children build the understanding and skills they need to succeed in school and beyond.

Homeschoolers, afterschoolers, unschoolers, and even classroom teachers appreciate our flexible approach that can work alongside any math curriculum.

Visit us today:
TabletopAcademyPress.com

Or browse Denise's blog:
DeniseGaskins.com

www.ingramcontent.com/pod-product-compliance
Lightning Source LLC
Chambersburg PA
CBHW061148070526
44584CB00034B/4454